RAND

Total Force Pilot Requirements and Management

An Executive Summary

Harry J. Thie, William W. Taylor,
Claire Mitchell Levy, Sheila Nataraj Kirby,
Clifford Graf II

Prepared for the
Office of the Secretary of Defense

National Defense Research Institute

Preface

This research was prompted by projections of pilot shortages in one military service even as that service was coping with existing pilot overages during the drawdown. Department of Defense policymakers became interested in whether future shortages would be a more widespread problem: would it affect more than one service, would it affect active and reserve components, and what might be done to resolve it. The purpose of this research is to examine the supply and demand for pilots and to address these questions.

This report summarizes the results of research first presented to members of the Office of the Secretary of Defense in November 1992. Updated assessments were presented in March, July, and November 1993, as results of the research became available. One report (Thie et al., 1994) presented results of the research on supply and demand through 1994. Another report (Levy, 1995) analyzed the interactions of military pilot supply and demand with civilian airline hiring. These reports should be of interest to military manpower planners across all services.

This research was sponsored by the Under Secretary of Defense for Personnel and Readiness and the Assistant Secretary of Defense for Reserve Affairs and was undertaken within the Defense Manpower Research Center, part of RAND's National Defense Research Institute, a federally funded research and development center sponsored by the Office of the Secretary of Defense, the Joint Staff, and the defense agencies.

Contents

Figures

Tables

Acknowledgments

The authors wish to thank the staff of the Under Secretary of Defense (Personnel and Readiness) for their support, particularly Dan Gardner, Wayne Spruell, and Mike Parmentier. We were helped throughout the course of this research project by many other officers and civilians in the Defense, Air Force, and Navy departments, and we owe them our thanks as well. A RAND colleague, Ron Sortor, reviewed the drafts of all the publications from this research effort, and we are grateful for his invaluable advice.

1. Introduction

The services are undergoing a fundamental reshaping and restructuring, driven by the demands of an era of tighter fiscal constraints, new security challenges, new technology, and increased reliance on the reserve component. To position themselves in this new environment and for drawing down, the services are constraining accessions, encouraging voluntary departures, and imposing involuntary separations. There is considerable uncertainty as to what these actions bode for the future, in terms of supply and demand of personnel and the sustainability of the required force.

This report focuses on one particular group of officers, military pilots. It assesses the ability to sustain the required pilot force and the personnel management tools available to manage the flow of pilots.

Rationale for and Background of the Study

The primary rationale for the study came out of the services' FY93 request for Aviator Continuation Pay (ACP) bonuses for pilots—a compensation package designed to improve retention of military pilots in certain critical areas. Despite the current environment of pilot *overages,* in its FY93 ACP submissions, the Air Force projected a future *shortage* of pilots, beginning in FY93 and widening significantly by FY96. Faced with this seeming contradiction, the Assistant Secretary of Defense (Force Management and Personnel) and the Assistant Secretary of Defense (Reserve Affairs) asked RAND to undertake a comprehensive assessment of issues related to pilot management and training.

The project focused on the following questions:

- What are the historical personnel trends for pilots in accessions, retention, and transfer rates between the active and reserve forces?

- What are the current requirements for pilots? How are they changing and why?

- What is the interaction between the military and civilian demand for pilots? What effect will this have on the sustainability of both the military and civilian pilot force?

2

- What are the current personnel and training policies used to manage pilot accessions and retention and how effective will they be in light of changing requirements for pilots?

The priorities for analysis were to examine, in order, the active Air Force, Air Reserve Component (ARC), Navy, and Naval Air Reserve.

The analysis was completed in 1994 and the results of the analysis were documented in two reports: (1) a report on the requirements for and supply of pilots in the context of the defense drawdown and restructuring, given historical trends in accession, retention, and transfers to the reserves, and the effectiveness of current personnel management and training policies in meeting future needs (Thie et al., 1994); and (2) a report on the interaction between the civilian airline industry and the military, which assessed the adequacy of future flows from the military to the civilian airlines (Levy, 1995).

After the research was completed, several issues surfaced because of our analysis and conclusions and other RAND work. The data we had used were necessarily somewhat dated and the restructuring that the services were undergoing was not fully reflected in the data. As more recent data became available, we questioned the validity of the analysis and conclusions in light of the new data. Second, our analysis pointed to the importance of assessing sustainability of both the active and reserve components conjointly, not in isolation. Third, a congressionally mandated RAND study of future officer careers presented a number of new career-management alternatives; we examined the effect of these alternatives on the training and retention of the pilot force (NDRI, 1994).

As a result, we were asked to undertake a short, follow-on study to:

- Update our inventory and requirements analysis with more current data and reassess our earlier conclusions regarding pilot shortages;

- Assess the adequacy of the flow of prior service pilots to the Selected Reserve components and suggest options to mitigate potential shortfalls in manning the reserves;

- Explore how the future experience profile of the pilot force might change and the implications of such a change;

- Review the recommendations of the concurrent study examining future career management for military officers in terms of their implications for managing the pilot force in the future.

This executive summary presents the results of both the original and the follow-on studies, although the primary emphasis is on the latter.

Historical Overview

To set the current situation in context, it is helpful to review some historical trends in pilot requirements, supply, and management. Figure 1 presents a historical look at the Air Force. As the figure makes clear, there has been constant change over the past forty years, both up and down. Overall, the long-term trend for pilots is down. In general, requirements appear to change more quickly at the beginning of a period of growth or decline, but inventory appears to lag behind. The services cannot develop or separate military pilots as quickly as their needed numbers can be changed on paper. During military buildups, the inventory of pilots typically falls short of requirements. This was the case during the Korean War of the 1950s, the Vietnam conflict in the 1960s, and the Reagan buildup of the 1980s. As forces are reduced during postconflict eras, inventory tends to exceed requirements.

Recognizing that supply changes will always lag requirements changes, the policy question is how to bring inventory and requirements closer to equilibrium. There have been two traditional long-term policy "balancers." First, the accessions and training of new pilots can be increased or decreased. However, when one compares the Air Force's pilot requirements with Air Force

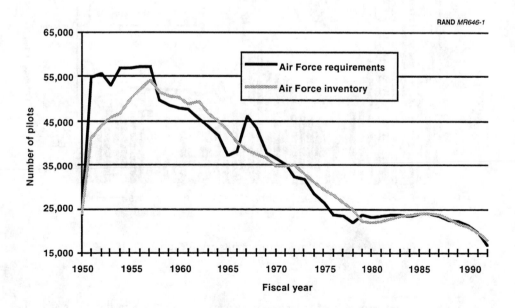

Figure 1—Air Force Pilots, 1950–1992

4

Undergraduate Pilot Training (UPT) accessions over a forty-year period (Figure 2), it is clear that changes in requirements are not immediately met by changes in training, because commitments to UPT are typically made to and by undergraduate students several years before they begin training. For example, there was a dramatic reduction in the requirements for pilots in 1967, as the Vietnam War began to draw down. However, reductions in UPT accessions did not begin until 1972.

A second long-term policy "balancer" has been retention. The measure of effectiveness used for retention by the Air Force is the cumulative continuation rate (CCR),[1] a measure of retention in the 6–11 year of service (YOS) group (the critical time after the active-duty service obligation ends and before officers make career decisions to stay until retirement). This measure has existed only since 1976, when the Air Force developed its rated management process.

In the short run, the Air Force has traditionally used the assignment system to mitigate shortages and surpluses. A priority assignment system, in which certain requirements are filled on a priority basis with some other requirements left unfilled, has been used to cope with historical shortages of pilots. During the

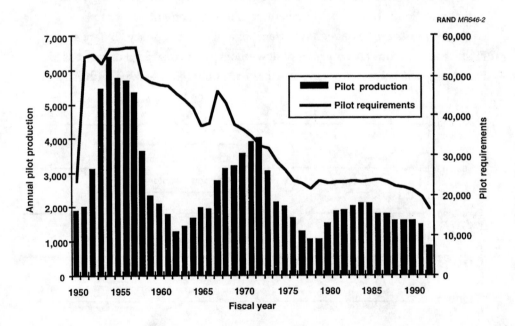

Figure 2—Air Force Pilot Requirements and Production, 1950–1992

---

[1]CCRs represent the product of multiple-year retention rates. CCR is normally calculated by the Air Force using the 6–11 YOS group and represents an estimate of the percentage of officers entering their sixth year of service who, given current retention patterns, are expected to remain in the service through their 11th year.

surplus period in the 1980s, the "rated supplement" allowed the assignment of rated officers not needed for pilot assignments to nonrated career fields for career development. This also provided a surge capability to meet contingency requirements in the past.

The Navy uses the active-duty reserve to help balance inventories and requirements. Increasing pilot requirements are met by increasing the number of newly commissioned U.S. Naval Reserve (USNR) officers to enter flight training. The number and proportion of active-duty reserve pilots declined from the late 1960s and early 1970s to the early 1980s. In 1971–72, approximately 35 percent of pilots were active-duty reservists. However, as requirements declined, these pilots were released from or not brought on active duty. By 1980, only 13 percent of pilots were active-duty reservists. This proportion rose during the Reagan buildup of 1980 to 1988, when active-duty reservists constituted about 30 percent of pilots. As of 1994, 23 percent of pilots are active-duty reservists and this number is expected to continue to decrease. To accommodate the ongoing drawdown, the Navy is separating USNR pilots who would be willing to stay.

Current Context

We are currently in the midst of a period of dramatic change, largely driven by the post–Cold War drawdown. Forces are being reduced significantly, first to a base force level, and then to a Bottom-Up Review level by 1999 (Aspin, 1993). The aviation drawdown has occurred more rapidly in the Air Force than in the Navy. By 1992, the Air Force had absorbed approximately two-thirds of the base force cuts. The Air Force (and to a lesser extent the Navy) also began a major reorganization in 1992, which affected all levels from Air Staff to the wing. In the Navy, squadrons and air wings will differ from their predecessors in their aircraft mix and number of aircraft. Marine Air will be more closely integrated with Naval Air. Finally, force structure shifts between active and reserve components will have important effects on requirements.

Our research addresses these issues and discusses new policies needed to deal with this changed and still changing environment.

2. Pilot Requirements and Inventories

Determinants of Pilot Requirements

We were asked first to examine why requirements have changed, specifically, what underlying factors have driven those changes, and then to look at how pilot requirements have changed since 1988 (the date of the DoD Aviator Retention Study, which was the last comprehensive work on aviator supply and demand issues). Third, we were to project how pilot requirements would change in the near future.

We developed three classes of determinants—force structure, organizational structure, and other policy change. Force structure changes center on the drawdown to include fewer primary aircraft authorized (PAA). Organizational restructuring covers the full spectrum of internal reorganization that the services are undergoing. In addition, we examined other policy changes that affect pilot requirements, such as changes in crew ratios.

We adapted Air Force terminology and divided pilot requirements into four basic categories—force, staff, training, and manyear or other, which includes transients and students. Force requirements include all of the pilots required to man operational flying units. Training programs are necessary to ensure the operational readiness and skill level of the operational units, including both formal and other training. Staff are necessary to provide overhead support and supervision to ensure safe, efficient, and productive flying operations. Finally, manyear or other requirements account for pilots in training, pilots participating in professional development or formal education programs, or pilots taking leave in conjunction with reassignment.

Determinants of Pilot Inventory

Pilot inventories, regardless of service, are determined by (1) production and (2) retention. Inventories are also influenced by other factors to include: (3) distribution; (4) absorption capacity; and (5) assignment and promotion policies. We briefly define these terms below.

Production

The annual rate of graduates from UPT. This factor is closely related to *UPT accessions;* the differences are due to attrition during UPT.

Retention

A measure of the likelihood that pilots will remain in the pilot inventory. This factor is determined by *loss rates,* which give the percentage of a cohort lost from the inventory in a given year. Forced retention has been legislated in the form of a minimum active-duty service obligation (ADSO) following the completion of flying training (currently eight years). Voluntary retention can be influenced by promotion, compensation, quality of life, and other factors such as civilian airline job opportunities and opportunities in the reserve component. Officers can also be promoted out of the pilot inventory, which includes only grades through O-5.

Distribution

The assignment of new pilots (and others not identified for major weapon systems (MWS)) to actual pipeline training that establishes MWS credentials.

Absorption Ability

The number of new and previously qualified pilots that each MWS group can accept each year. This factor is influenced (or constrained) by several policy issues and other parameters. These include:

Experience. This parameter has two components. The first establishes criteria on minimum flying time and/or time in crew position required for a crewmember to be identified as experienced in a given MWS. The second determines minimum proportions of aircrew authorizations (by flying unit) that must be filled by crewmembers who meet these criteria and are thus identified as experienced.

Stability. This includes (1) permanent change of station (PCS) and assignment stability—length of time at one base, unit, or position before PCS move; (2) weapon system stability—length of time flying a particular weapon system; (3) aircraft commander stability—length of time in aircraft after upgrade to aircraft commander; (4) PAA and crew ratios (by MWS)—all absorbing cockpits are part of the crew ratio force; (5) vacancies in absorbing units, or the number of *absorbing cockpits*; (6) pipeline (i.e. post-UPT) training capacity (by MWS); (7) loss rates and retention (which are, in turn, affected by a number of other factors).

Absorption ability clearly influences (and constrains) the distribution and number of new pilots to be produced each year.

Interaction of the Factors

These factors dynamically interact to complicate real-world inventory management. For example, over time the Air Force has moved from accessing pilots to sustainment levels, to accessing pilots to absorption ability, and finally to a vacancy-driven system as the factors have changed.

Civilian Airline Hiring

Retention is a key factor affecting pilot inventory. A central problem perceived by some for retention of military pilots is civilian airline hiring.

Our analysis shows that in the near term, there are enough separated military (and other) pilots to satisfy civilian airline hiring requirements through the end of the drawdown in 1997. On the demand side, the airline industry will continue to struggle through the end of this decade, with depressed hiring. On the supply side, a large number of pilots will be separating from the military and will be available for civilian hiring. They will constitute current losses and losses from previous years who have not been able to find jobs in the civilian sector.

After the year 2002, airline hiring will increase again at a modest pace, fueled at least partly by the need to replace aging Vietnam veterans and partly by projections of increased, though still moderate, growth in the airline industry. During this time period, the available pool for hires will become much smaller because of the markedly smaller cohorts of military pilots being accessed currently. This imbalance may well cause problems for military pilot retention, unless civilian airlines can "grow their own pilots."[1]

Air Force Inventory Compared to Requirements

An examination of the underlying determinants that drive changes in pilot requirements provides a window into what these future requirements are likely

---

[1]A more important problem, from the military perspective, is the inadequacy of the flow of military pilots to the Air Reserve Components—a problem discussed below.

10

to be. Force requirements, which are driven by the crew ratio subcategory, are a direct function of force structure (PAA times crew ratio times crew complement). This category is reduced by 34 percent, which is similar to the cut in PAA in the base force. Earlier data regarding staff requirements did not appear to fully reflect the ongoing Air Force reorganization. In particular, wing and below-wing staff reductions of 15 percent did not reflect the projected 20 percent reductions from the new objective wing nor the 27 percent reduction in the number of wings.

Our earlier analysis projected that staff requirements would be reduced by a total of 20 percent when all organizational changes were fully reflected in databases. As a result, our earlier work estimated that FY93 and later requirements would be lower than original Air Force projections by about 14 percent—10 percent in further force reductions[2] and 20 percent in staff requirements reflecting decisions already made about reorganization. Using the Air Force inventory line, as shown in Figure 3, we concluded that there should be no near-term problem in meeting pilot requirements, although we warned that some of the changes in requirements could exacerbate or cause long-term problems. Since our initial

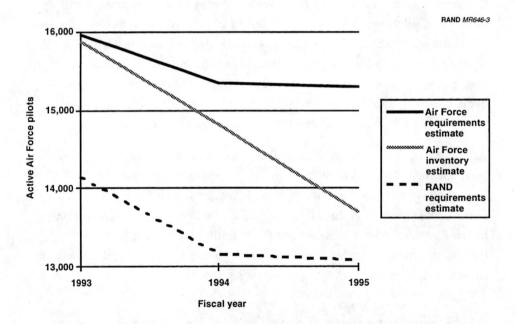

Figure 3—Projected Near-Term Requirements for Air Force Pilots Under Air Force and RAND Assumptions

---

[2]At the time of the requirements work, the base force was the official plan for force cuts. We postulated that there would be an *additional* 10 percent force cut, on top of the base force, as called for in a plan put forth by Congressman Les Aspin. Subsequent force cuts in the *Bottom-Up Review* have come close to matching our projection of a further 10 percent force cut.

work, the Air Force has revised downward its estimate of requirements in the near future as shown in Figure 4; this estimate reflects force cuts called for in the *Bottom-Up Review* (Aspin, 1993), and a further 20 percent decline primarily in staff positions, much as we had predicted. These cuts in staff have been assessed to the major commands and presumably will take place in the near future. The new requirements line is close to the RAND estimated line shown in Figure 3. Clearly, this new requirements line is predicated on the current force structure; if the force structure changes or if greater staff efficiencies could be found, this line could come down still further. Nonetheless, the remainder of the analysis is based on the Air Force requirements projections.

Since we ultimately want to compare future inventories of pilots with requirements for them, we used the Air Force pilot requirements to estimate objective profiles for those requirements in the steady state (FY97+). These profiles convert data by grade to data by years or length of service. We use them as an objective that policies for pilot management and training are trying to achieve. Such policies change inventory profiles, which can then be compared to the requirements profiles to see whether and if the objectives are being met.

We first projected a future FY97 inventory line.[3] This was then matched to the projected FY97 requirements line to see whether problems of mismatch are likely

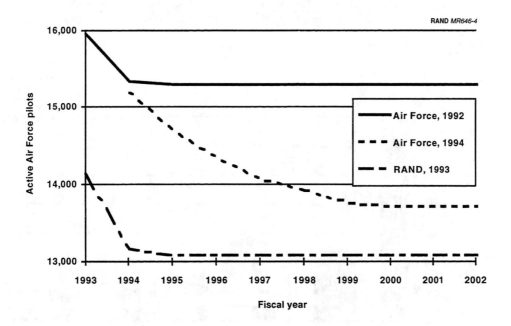

Figure 4—Changes in Air Force Requirements

---

[3]See Thie et al. (1994) for a more detailed discussion of our projections.

to exist in the near future. Our basic conclusion, shown in Figure 5, is that no critical numerical shortages will exist in the aggregate or in any major weapon systems through FY97.

Although there is no short-term numerical shortage of pilots, there is a serious maldistribution of experience when we compare the FY97 inventory projection for a given year of service with the corresponding requirements objective profile for that year. This maldistribution generates no *immediate* operational problem, since shortages occur among the least-experienced (and presumable least-capable) pilots, and the overages occur among the most-experienced cohorts. However, an important question is: What is the potential operational effect as the shorted cohorts age to where they would normally assume increased flying responsibilities? To examine this issue, we continued our projection model for another five years.

Figure 6 confirms that the damaged cohorts will generate a "bathtub" effect in the inventory by FY02, and it also establishes that significant inventory shortfalls will occur by that time. On the basis of early 1995 planned training levels, we estimate that the FY02 inventory will consist of 12,300 pilots and the Air Force projects a requirement of 13,700 pilots. The shortage is particularly distributed in the 6–12 YOS group—the critical years for having experienced pilots in cockpits in operational units. This bathtub is likely to cause serious problems by FY02 and later. In general, individual MWS reflect similar patterns to those exhibited by the total inventory.

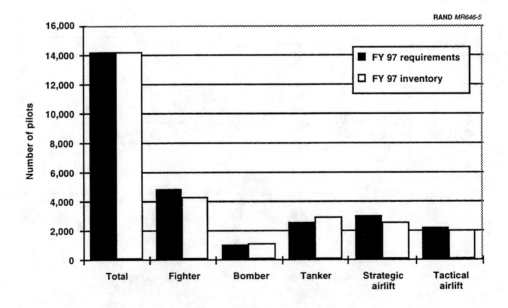

Figure 5—Projected Supply of and Demand for Air Force Pilots in FY97

To place the overall estimates of expected overages and shortages in context, we reproduce the earlier chart and compare our estimates with historical swings in pilot inventory and requirements (Figure 7).

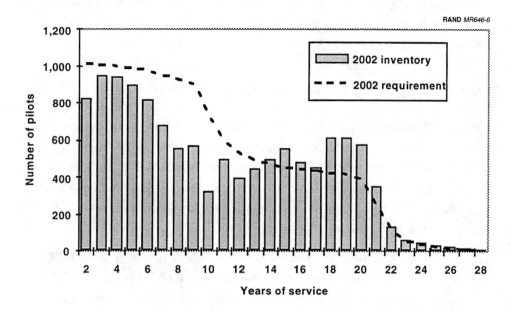

Figure 6—Projected YOS Profile of Air Force Pilot Inventory and Requirements in FY02

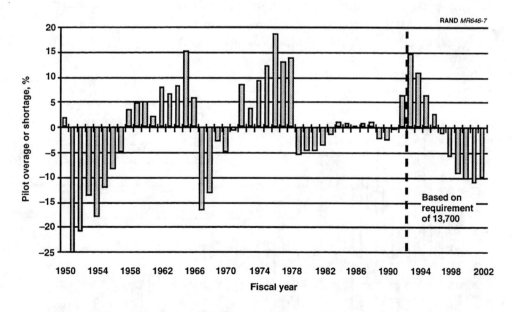

Figure 7—Historical Air Force Pilot Overage and Shortage Proportions (computed as inventory minus requirement divided by requirement)

There is a long-term downward trend in the magnitude of fluctuations—the overages experienced in FY91–FY95 are clearly not as large as we experienced during the late 1970s and the shortages projected for the outyears are not as large as the nation experienced during the 1950s or at the end of the Vietnam War. However, it is clear that the near term will undoubtedly experience the same swings that have been observed historically.

Navy Inventory Compared to Requirements

Specific details about the Navy inventory and requirements lines are provided in our earlier report. Our basic conclusion is summarized in Figure 8—the Navy, unlike the Air Force, will not face any major problems in this area in the future. (Although not shown here, this conclusion holds for the near term as well, where the projected FY97 requirement line, matched against the FY97 inventory line, shows no major experience maldistribution either.)

Alignment of inventory and requirement distributions for the Navy in FY02 is remarkably close. Indeed, an overall 200-pilot overage is a significant improvement over the pre-drawdown situation in FY92, which showed a shortage of 2,000 pilots. (Some drawdown losses probably occurred in FY92 inventory that were not reflected in requirements.) There are several ways to obtain the slight retention improvements required to match the requirement numbers, and it is also possible that requirements will continue to drop in the outyears if the Navy initiates aggressive policy action.

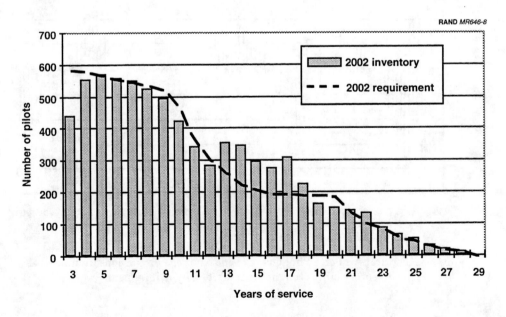

Figure 8—Projected YOS Profile of Navy Pilot Inventory and Requirements in FY02

Differences Between Air Force and Navy Comparisons

We can offer two reasons to explain the marked difference in the projected outcomes for the Navy and the Air Force. First, somewhat ironically, the Navy's higher attrition experience before ten years of service (due to shorter active-duty service obligations and lower natural retention) allows the Navy to avoid the significant overages encountered by the Air Force in the YOS 7 through YOS 12 cohorts and enables the Navy to maintain reasonable UPT production levels throughout the drawdown period, which will help avoid future cohort shortfalls. Second, the Air Force began its drawdown sooner than the Navy and aggressively implemented its plans. Although some experienced pilots were cut, the Air Force cut training of new pilots significantly in the face of an inability to place newly trained pilots into cockpits. This measure leads to the maldistribution of experience—or what we referred to as the bathtub effect—as the "damaged" cohorts fall short of the numbers required in the 6–12 YOS experience groups in later years.

The conclusions reported here are somewhat conditional, because the potential severity of the problems will depend critically on the policy options chosen to counter the problems. Part of our tasking was to examine available options and how successful they might be in alleviating these problems. Although we focus on the Air Force for the remainder of our analysis, many of the policy options, of course, apply to the Navy.

3. Air Force Pilot Management and Training

If we are projecting a shortfall of pilots in the future, the two policy choices that we need to look at are training larger numbers of new pilots and retaining more experienced pilots. These are the two key determinants of pilot inventory. The discussion below outlines several alternative policy options and examines the effect of each on the problems identified in the previous section.

Effects of Changed Retention Policies

The first two alternatives we examine are based on changed retention. To build what could be considered a "control" or comparison group, we first characterized a base case. Retention is measured in terms of Total Active Rated Service (or TARS), which is the expected number of years of rated service that an average pilot will serve on active duty after completing UPT.[1] Future retention behavior was estimated by adjusting historical retention data for changing circumstances. For example, one significant change is the increase in ADSO incurred by UPT graduates, which created a retention point at eight years of rated service (approximately 10 years of active service).[2] For the base case, we used the following TARS values for pilots who would reach the retention point in the years indicated:

1. FY95–FY97 (high[3] retention):   TARS = 13.5 years;

2. FY98–FY99 (lowest retention):   TARS = 12.0 years;

3. FY00–FY01 (low retention):   TARS = 12.5 years.

---

[1]TARS values have historically ranged from 11.1 to 14.1 years.

[2]The retention point is based on years of *rated* service, not the normal year-of-service unit that determines promotion or retirement eligibility and that is used in the various figures in this section. The retention point typically occurs at 10 YOS (due to the length of UPT), but it can come at 11 or 12 YOS (or even later) for officers who are delayed in starting (or finishing) UPT. We used historical data to estimate the year-of-service distribution for UPT graduates. Expectation of behavior at the retention decision point also had to be modified. Previously, the retention decision could be made by a pilot over a period of several years following end-ADSO (at six years of rated service). An Air Force drawdown initiative now requires eligible pilots (i.e., fixed-wing pilots in good standing at end-ADSO) to take the pilot bonus (and incur an active-duty commitment until 15 YOS) or accept a nonflying assignment outside the aviation career track.

[3]The modifiers—high, lowest, low—refer to the retention rates (TARS slightly over 13 years) that we would expect without making the additional adjustments described in the rationale.

18

The rationale for these assumptions is discussed below.

Pilots reaching the retention point in FY95–FY97 are in cohorts that have been "shaped" by the drawdown and the officers remaining in these cohorts should be a select group. Thus, it seems credible to assume that when they approach the retention window, they should have higher retention than might typically be the case. Figure 9 illustrates what we mean by shaped cohorts. The dotted line shows the FY93 year of service requirements profile matched against the FY93 beginning-year inventory, shown by the full height of the bars. Because of the large overages in the more-experienced year groups, the Air Force instituted a series of force reduction policies targeted at these groups. For example, the Pilot Early Release Program (PERP) was targeted at those in the 6–14 YOS groups, Variable Separation Incentives/Selective Separation Bonuses (VSI/SSB) were offered primarily to those in 14+ YOS groups, and Selected Early Retirement Boards (SERB) were mandated for some in the 24+ YOS groups. As a result of the reductions (shown by the lighter or darker areas at the top of the bars), the FY93 end-year inventory more closely matched the required profile. This provides the rationale for the assumption of high retention for these cohorts in the base case described above. Many of those who might have left later have already done so earlier.

The number of pilots reaching the retention point in FY98–FY99 is reasonably large (especially when viewed relative to requirements for these cohorts),

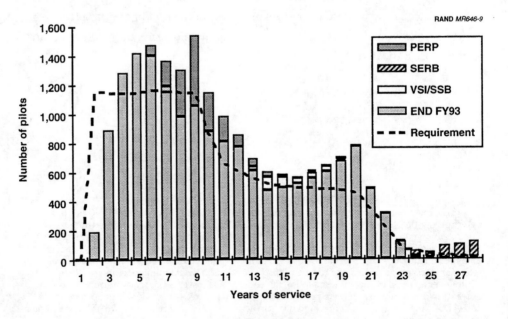

Figure 9—Example of Shaping Cohorts

whereas those at the retention point in FY00–FY01 are smaller because fewer pilots were trained in the early 1990s; the generally observed inverse relationship between relative cohort size and retention provides the rationale for the particular assumptions we adopt for these cohorts.

Table 1 outlines Alternatives 1 and 2 (as well as the base case for contrast), which bound the realm of possibilities with respect to changed retention. Alternative 1 assumes uniformly high retention for all cohorts, regardless of size or year (13.5 TARS); Alternative 2 assumes a much lower retention for the shaped cohorts (12.5 instead of 13.5 TARS), while maintaining the assumptions regarding retention of the cohorts in FY98–FY01 (12.0 and 12.5 TARS, respectively).

The results of our analysis under the different alternatives are displayed in Figure 10. The year of service requirement profile is shown by the dotted line, and the dark grey bars depict the FY02 inventory under the base case. Both of these are repeated from Figure 6. Recall that we estimate a shortfall in the base case of 1,400 pilots. The bathtub is very evident. What is interesting and perhaps surprising is that changed retention appears to only marginally affect the inventory profile. In the aggregate, we estimate that increased retention (Alternative 1) will increase the overall inventory to 12,900, thus reducing the estimated shortfall by 600 pilots, whereas lower retention (Alternative 2) will decrease the overall inventory to 11,700, thereby increasing the shortfall by 400 pilots. Note that there is, however, no measurable effect on the maldistribution of experience in the early years of service. The different retention scenarios affect primarily those in the more-experienced year groups, 10–18 YOS, with little or no effect on the bathtub. This is because UPT production was held fixed at projected Air Force levels (through FY01) for these excursions. Thus, training could not respond to retention changes until FY02.

Table 1

Assumptions Underlying Base Case and Retention Alternatives

Year	Base Case	Alternative 1: High Retention	Alternative 2: Low Retention
FY95	13.5	13.5	12.5
FY96	13.5	13.5	12.5
FY97	13.5	13.5	12.5
FY98	12.0	13.5	12.0
FY99	12.0	13.5	12.0
FY00	12.5	13.5	12.5
FY01	12.5	13.5	12.5

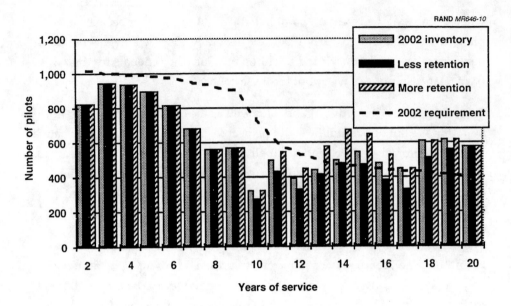

Figure 10—Effects of Retention Alternatives

Effects of Changed Training Policies

The next alternatives focus on increased training. The base case is drawn from Air Force plans that were in effect at the time of the analysis. After that, the Air Force, recognizing the future value of increased UPT now, revised planned training rates upward to those shown in the second column of Table 2.[4] As a result, the base case shown does not fully mirror current Air Force estimates. We examine two alternatives: an increase in UPT of 100 starting in FY00; and an increase in UPT of 100 starting in FY97 and continuing each year thereafter. The

Table 2

Assumptions Underlying Base Case and Training Alternatives

Year	Base Case	Revised Air Force Projections	Alternative 3: Increase UPT Later	Alternative 4: Increase UPT Earlier
FY95	500			
FY96	525			
FY97	670	700		770
FY98	811	925		911
FY99	911	950		1,011
FY00	990	1,025	1,090	1,090
FY01	1,050	1,050	1,150	1,150

[4]If these new training levels are implemented, their effect is reasonably shown by Alternative 4.

numbers selected were premised on feasibility and reasonable access to training bases and training aircraft.[5]

Figure 11 shows the results of increased UPT relative to the base case.[6] Increasing UPT earlier offers the best solution thus far, in terms of both increased inventory (+500 over the base case) as well as significant effects on the bathtub. Indeed, the effects on the 5–7 YOS groups is quite pronounced and the inventory profile for these groups appears to be remarkably close to the requirements line. Increasing UPT later offers a minor inventory increase of 200 pilots, with more modest effects on alleviating the bathtub effect.

Effect of Changed Training and Changed Retention Policies

We explored one last alternative that combined both changed retention and increased training, as shown in Table 3. We thought it might be interesting to consider a real-world situation in which the shaped cohorts had lower retention (as envisaged in Alternative 2), but the Air Force, in response to the much lower

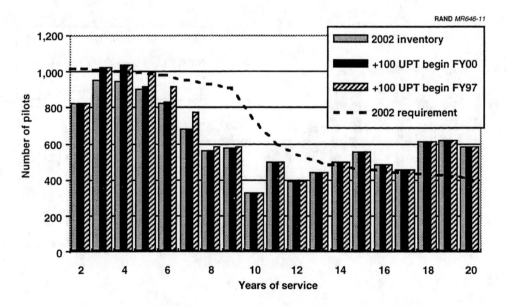

Figure 11—Effects of Increased Training

---

[5]During the 1980s, UPT training was typically 1,500 to 2,000 per year.

[6]Some years of service exhibit the accumulated effect of UPT graduates entering the pilot inventory with three or more years of service.

22

Table 3

Assumptions Underlying Base Case and Retention/Training Alternative

Cohort Year	Base Case Retention Assumptions	Alternative 5 Low Retention	Base Case Training Assumptions	Alternative 5 Increase UPT After Lower Retention Is Known
FY95	13.5	12.5	500	
FY96	13.5	12.5	525	
FY97	13.5	12.5	670	720
FY98	12.0	12.0	811	911
FY99	12.0	12.0	911	1,011
FY00	12.5	12.5	990	1,090
FY01	12.5	12.5	1,050	1,150

than expected retention, increased UPT training by 100 for each year (except the first year, FY97).[7]

The result of this scenario on the FY02 inventory is presented in Figure 12. The combined effect of low retention and increased training is to increase the estimated pilot shortfall by 100. Increased training offsets to some extent the small negative effects of reduced retention because it allows the service to gain

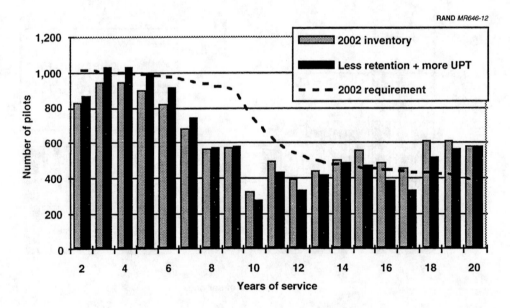

Figure 12—Effect of Lower Retention and Increased Training

---

[7]Our reasoning was that by the time the Air Force could recognize the lower retention rate and act on it, it would not be feasible to increase training by 100; hence we adopted the lower number for FY97.

back those lost through lower retention and, in addition, mitigates the bathtub problem as well because the pilots are distributed better in terms of required experience.

Conclusions

No likely policy scenario affecting inventory alone will completely resolve the twin problems of overall shortage in the outyears and maldistribution of experience. Figure 13 summarizes our assessment of the base case and the five scenarios that we considered: Alternative 1 (low retention), Alternative 2 (high retention), Alternative 3 (increased UPT later), Alternative 4 (increased UPT earlier), and Alternative 5 (low retention/increased UPT earlier). The low retention scenario offers the worst-case scenario: The shortfall increases relative to the base case and the bathtub effect worsens. Increasing UPT earlier—the policy largely adopted by the Air Force in their current plans—offers the best solution: The shortfall declines and there is less mismatch between the required and actual experience profile. Increased retention has the biggest positive effect in reducing the magnitude of the expected shortage, but it does not significantly affect the bathtub and increases the overage in later years of service.

The Air Force will likely face overall shortfalls of pilots by FY02 and, perhaps more serious, a significant maldistribution of experience. Changing retention

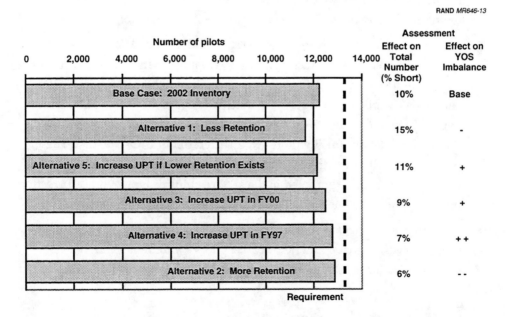

Figure 13—Assessment Summary

24

does not offer much promise for changing either of these two outcomes, but reducing requirements and increasing training do. Reducing requirements even beyond the current cuts planned by the Air Force, particularly in the staff category, is crucial to reducing future shortfalls and increasing UPT is fundamental to solving the bathtub problem. However, the Air Force needs to be able to absorb graduates of UPT in cockpits and operational units, and assignment policy is the key to increasing absorption. As such, a fundamental reexamination of assignment policy to accommodate these new goals seems warranted. Indeed, the Air Force is already doing this by moving away from the voluntary assignment system into one better suited to the needs of the service in terms of flows of individuals.[8]

However, a number of measures can be taken in the short term to minimize the effects of the bathtub and shortage problems. We list these below.

1. *Active-duty tours for certain ARC pilots* would directly increase active inventory. The Air Force is currently employing a policy similar to this in a modest way.[9]

2. *Reserve/civilian instructor pilot manning* would directly decrease active-duty pilot requirements, although questions of experience and training might arise.

3. *Changed assignment policy* would accelerate absorption of new pilots into units as more experienced pilots move into staff or other positions, opening up the cockpits to new pilots. This would increase inventory.

4. *First assignment instructor pilot (FAIP) manning* decreases MWS-qualified specific requirements because new pilots remain as instructors after finishing UPT to fill nonspecific MWS positions. In addition, this allows the FAIPs later to be absorbed into units more quickly because they have more flying hours. Throughout the 1980s, the Air Force employed this policy but then dropped it in favor of using more-experienced pilots as instructors. These pilots, it was felt, brought with them actual experience in operational units and were more strongly rooted in the service culture. However, resurrecting the FAIPs might be one way of decreasing requirements.

5. *Prioritizing assignments for fill* (as has been done historically) is yet another stopgap measure that could help reduce pilot requirements and, hence, expected shortages. Pilots could be used to fill the top-priority positions and

---

[8]See Thie et al. (1994) for a more detailed discussion.

[9]The Air Force recently announced a voluntary pilot-recall program for 250 fighter pilots with recent active-duty or Air Reserve Component experience.

nonrated officers could be used to fill other positions or these positions could be left vacant.

In the next section, we discuss the second task of our overall project: an assessment of the adequacy of pilot flow to the reserves.

4. Pilot Flow to the Air Reserve Component

The post–Cold War drawdown is changing the force mix equation particularly for the Air Force,[1] with possible implications for pilot flows from the active to reserve components. These flows are critical to the Air Force Reserve (AFRES) and Air National Guard (ANG), which rely heavily on accessions of prior service pilots.

A growing share of aircraft and missions are being housed in the ARC as the Air Force draws down. Although the post–Cold War drawdown has reduced active Air Forces by roughly one-third, the Air Reserve Components have not faced significant cuts. ARC units are slated to grow from 33 percent to 42 percent of the total force by 1995. Active/reserve mix in several MWSs, including fighters, tankers, and tactical airlift, will be close to 50 percent.[2] This underscores the importance of keeping a total force perspective when analyzing the effects of reducing active force size.

Figure 14 distributes total 1989–1992 ARC hires by source: prior active service; prior reserve service, UPT, and other, which includes experienced pilots from other services. The ANG draws over 50 percent of its hires from the actives and the AFRES is even more heavily dependent on them. The median prior service hire into the ARC had completed 7 years of service (or the minimum ADSO) and approximately 85 percent of them have 6–12 YOS.

Typically, 40 percent of Air Force pilot separations affiliate with the ARC and there is anecdotal evidence to suggest that the ARC have been demand-constrained. Indeed, AFRES cites a waiting list of over 1,500 in 1993 and certainly, with the current AF drawdown, the ARC faces a rich recruiting environment with a larger-than-usual reserve pool of prior service separations on which to draw. This is illustrated in Figure 15, which adopts a loss cohort perspective. It shows Air Force losses by year and the cumulative number of ARC hires from each of these loss cohorts.

---

[1]The Navy should not have a problem manning its reserve pilot force. The active-reserve ratio (in terms of tactical aircraft) is around 6 for the Navy and increases to between 10 and 12 in later years (FY95–FY97) because of the loss of an air wing in the reserve.

[2]For FY96, tactical airlift, aerial refueling (strategic tankers), and tactical fighter units of the Air Reserve Component aircraft are 78 percent, 56 percent, and 40 percent of the Total Air Force (DoD, 1995).

28

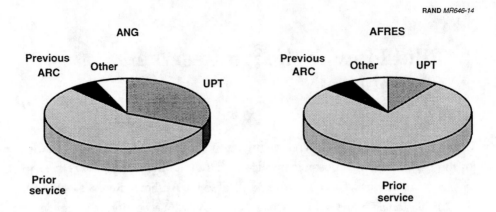

Figure 14—ARC Hiring

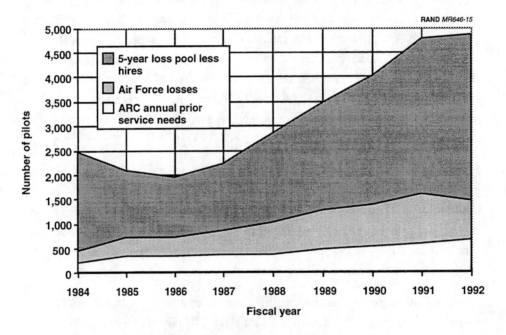

Figure 15—ARC Hiring, FY84–FY92

However, the situation is likely to change quite dramatically in the future as projected hiring requirements exceed projected losses from the active, as shown in Figure 16. ARC requirements remain fairly constant but losses (estimated using base case assumptions discussed above) from the Air Force will be considerably smaller because of the smaller cohorts. The situation may be somewhat mitigated by the large reserve pools that currently exist, but the ARC prefers to hire pilots within five years of separation, since these pilots require little or no retraining. Those separated for five or more years are typically regarded as ineligible or undesirable.

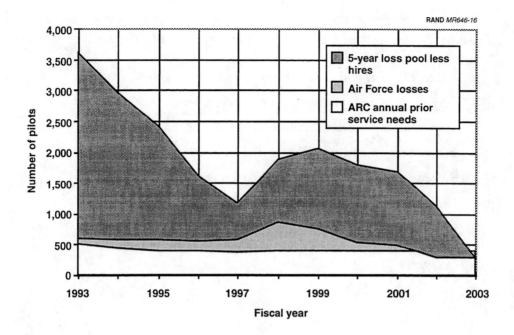

Figure 16—ARC Hiring, FY93–FY02

In particular, some MWSs in the ARC face a bleak and uncertain hiring environment, in which ARC annual prior service needs far outpace projected Air Force losses. For example, in tactical airlift, a likely future shortage is being driven by the very low numbers of such tactical air forces housed in the active compared to the reserve.

The discussion above underscores the need for a total force perspective when considering reductions in force for the active. Prior service pilots form the bulk of the experience and hires of the ARC pilot force and the reductions and policies being adopted by the active could seriously jeopardize the flow to the ARC and, thus, its future readiness. Closer reserve integration may well be the answer, modeled after programs such as the Army's 2+2+4 for enlisted personnel, which marries a reserve obligation to the active term of service.

Maintaining flows to the ARC should be a high priority for the Air Force and an integral part of their discussions when considering the implementation of policies aimed at pilot management. The large reserve pool of prior service pilots fueled by the active drawdown offers a breathing space before ARC readiness and recruiting are compromised and gives the services time to promulgate policies designed to forestall future problems in both the active and reserve components.

We suggest several policy options without any serious attempt at evaluation. Some may be infeasible or undesirable; all may warrant further study.

1. Increase UPT or find alternative sources for ARC pilots.

2. Retire certain ARC pilots later (e.g., strategic and tactical airlift pilots).

3. Coordinate ARC hiring within a five-year window.

4. Relax ARC ideal prior-service profile (in terms of YOS and affiliation gap).

5. Seek more prior-service pilots from other services.

6. Hire more prior-service ARC pilots.

7. Reduce ADSO and tie commitment to ARC service.

5. Other Issues

Changing Experience Levels of the Pilot Force

The third task we were asked to undertake was to explore how the overall experience level of the pilot force might change in the future and whether this should be a source of concern. We computed the experience ratio using pilots with less than 11 YOS as the denominator and those with 5–11 YOS as the numerator. In essence, this is a ratio of the more experienced pilots likely to be in cockpits to all pilots likely to be in cockpits. Figure 17 shows how this changes over time. The experience ratio just after Operation Desert Shield/Storm (ODS/S) is the starting point. The reduced training of new pilots in the Air Force sharply increases the experience ratio to what might be considered untenable levels. This gradually tapers off in the earlier part of the next century. However, the gradual reduction in overall experience is not a matter for concern: We merely see returns to ODS/S experience levels after an engineered and unsustainable increase brought on by reduced UPT levels. The Navy shows a

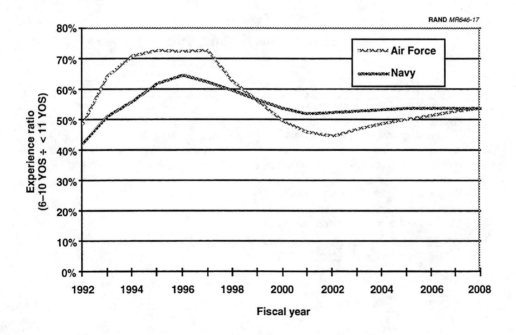

Figure 17—Experience Ratios, FY92–FY08

small increase during the late 1990s but gradually stabilizes to ODS/S experience levels after that.

Future Career Alternatives and Effects on the Pilot Force

The final task was to consider the effects on pilots' careers, flows, and experience of dramatically different career patterns (NDRI, 1994). Table 4 presents several career alternatives and our assessment of their effects on the need for UPT, experience, and flows to the ARC, relative to the status quo.

Longer active service careers would decrease UPT and increase experience but would reduce flows to the ARC. Higher continuation opportunities and selective promotion—such as fly-only careers in the active Air Force--would also have the same effects. Forced attrition to the ARC after a period of service would increase UPT to maintain the active force, would lower experience, and would obviously increase the flows to the ARC (its primary purpose). The effect of the earlier vesting but later retirement annuities alternative is much harder to predict because of changed incentives for staying or leaving. A final alternative considered in the study is that of more lateral entry from civilian life or from the reserves. This would enhance flows both to and from the reserves. Effects on UPT and overall experience are harder to model.

Table 4

Future Career Alternatives and Predicted Effects on Pilot Force

	Assessment		
Alternative	UPT	Experience	Flows
Longer active-service careers	Less	More	Fewer
Higher continuation opportunity and selective promotion (e.g., fly-only careers)	Less	More	Fewer
Forced attrition to ARC after period of active service	More	Less	Greater
Earlier vesting but later retirement annuities	?	?	?
More lateral entry (from civilian life or reserves)	?	?	Greater: two way

6. Conclusions

We summarize our main conclusions here:

1. The trend of pilot requirements should be downward with no major aberrations, although the active-reserve pilot mix is still uncertain.

2. Recent pilot undertraining in the Air Force will create future shortfalls and experience imbalances; further reductions in requirements and increased training are key parts of the solution. Short-term measures can minimize effects of shortfalls and imbalances.

3. Overall, smaller active inventories and proportionally larger reserves may result in insufficient future flow from active to reserve components. A policy of closer integration with reserve needs should be adopted.

4. New pilot career patterns need to be judged against future active and reserve needs.

References

Aspin, Les, *Bottom-Up-Review*, Department of Defense, 1993.

Department of Defense, *DoD Aviator Retention Study, Volume I, Study Report*, November 28, 1988.

Department of Defense, *FY 1996 Manpower Requirements Report*, March 1995.

Levy, Claire Mitchell, *The Civilian Airline Industry's Role in Military Pilot Retention: Beggarman or Thief?* Santa Monica, Calif.: RAND, DB-118-OSD, 1995.

National Defense Research Institute (NDRI), *Future Career Management Systems for U.S. Military Officers*, Santa Monica, Calif.: RAND, MR-470-OSD, 1994.

Thie, Harry J., William W. Taylor, Claire Mitchell Levy, Clifford M. Graf II, Sheila Nataraj Kirby, *A Critical Assessment of Total Force Pilot Requirements, Management, and Training*, Santa Monica, Calif.: RAND, DB-121-OSD, 1994.